# HOFLEHNER. JET AIRLINER. THE COMPLETE WORKS.

MOST/PRESS

# THE WORKS

2009-2011

The Jet Airliner series was taken over a period of several months between early 2009 and late 2011 at Maho Beach on the Dutch/French island of St. Maarten / St. Martin in the Caribbean Sea. The beach is directly adjacent to the relatively short runway (2,300m) of the airport. Therefore passenger jets roar as low as four meters above the heads of sunbathers, with speed of approx. 250 km/h. Most aircraft movements at the Princess Juliana International Airport account for small regional and private planes - with only 5 to 12 passenger jets arriving each day, depending on season. The Queen of the Skies, the Boeing 747, can be spotted just twice a week. For this project, two classic 6x6 Hasselblad wide-angle cameras were used (no autofocus, etc) - so only 1 shot per landing and camera was possible. Most important, the photographs are real and the situations with the people on the beach have not been arranged. All photographs show the actual live-action during a split second of a landing or take-off.

View of crowd at Maho Beach from window seat 09F

View from a US Airways 757, flight US1560 from Charlotte-Douglas
Cruising low above Caribbean waters, on short final to Princess Juliana Int'l Airport

PRINCESS JULIANA INTERNATIONAL AIRPORT
IATA: SXM / ICAO: TNCM

ST. MAARTEN / ST. MARTIN
18°02´27´´N / 063°06´34´´W / 14 ft

American Airlines Boeing 737-800
Flight AA687 from Miami International

American Airlines Boeing 737-800
Flight AA687 from Miami International

#01

#17

Delta Air Lines Boeing 737-700
Flight DL440 to Atlanta Hartsfield-Jackson

United Airlines Airbus A320-200
Flight UA39 from Chicago-O'Hare

#57

American Airlines Boeing 757-200
Flight AA667 from New York-John F. Kennedy

American Airlines Boeing 757-200
Flight AA687 from Miami International

#72

KLM Royal Dutch Airlines Boeing 747-400
Flight KL785 from Amsterdam-Schiphol

#20

US Airways Boeing 757-200
Flight US1560 from Charlotte-Douglas

Air Caraibes Airbus A330-300
Flight TX536 from Paris-Orly

TAKE-OFFS OVER BEACH ARE NOT THAT COMMON.

NOT EVERYBODY REALIZES WHERE THE EAR-BATTERING NOISE IS COMING FROM.

JetBlue Airways Airbus A320-200
Flight B6798 to Boston-Logan

#70

KLM Royal Dutch Airlines Boeing 747-400
Flight KL785 from Amsterdam-Schiphol

US Airways Boeing 757-200

Flight US1209 from Philadelphia International

#68

IN EARLY 2012, FRENCH LEISURE CARRIER CORSAIR HAS SUSPENDED SERVICE TO ST. MAARTEN.

THEIR JUMBO JETS ARE UNLIKELY TO RETURN.

Corsair Boeing 747-400
Flight SS868 from Paris-Orly

#28

US Airways Boeing 757-200

Flight US1209 from Philadelphia International

#23

#36

Continental Airlines Boeing 737-700
Flight CO861 from Newark-Liberty

Air Caraibes Airbus A330-200
Flight TX534 from Paris-Orly

#43

Continental Boeing 737-700

Flight CO818 from Newark-Liberty

Air France Airbus A340-300
Flight AF3510 from Paris-Charles de Gaulle

#06

American Airlines Boeing 737-800
Flight AA2160 to Miami International

Air France Airbus A340-300
Flight AF3510 from Paris-Charles de Gaulle

PHOTOGRAPH FROM 2009 THAT IS NO LONGER POSSIBLE TODAY.

TO COMPLY WITH ICAO RULES, THE RUNWAY, ON THE OPPOSITE SIDE OF MAHO BEACH, HAS BEEN LENGTHENED IN 2010.

ICAO = INTERNATIONAL CIVIL AVIATION ORGANIZATION

Air France Airbus A340-300
Flight AF489 to Paris-Charles de Gaulle

#81

US Airways Boeing 757-200

Flight US1562 from Charlotte-Douglas

Delta Air Lines Boeing 757-200
Flight DL439 from Atlanta-Hartsfield Jackson

POLICE OFFICERS ARE A REGULAR VISITOR.
THEY ALWAYS LOOK OUT FOR CARS THEY CAN HAVE TOWED-AWAY.

US Airways Boeing 757-200
Flight US1560 from Charlotte-Douglas

#52

#78

Air Caraibes Airbus A330-300
Flight TX536 from Paris-Orly

American Airlines Boeing 737-800
Flight AA687 from Miami International

#18

#66

Air France Airbus A340-300

Flight AF3510 from Paris-Charles de Gaulle

Air France Airbus A340-300
Flight AF3510 from Paris-Charles de Gaulle

#79

#54

Air Transat Airbus A310-300
Flight TS204 from Montreal-Trudeau

US Airways Boeing 757-200
Flight US1209 from Philadelphia International

#37

THE RAREST SIGHT ON ST. MAARTEN

KLM IS THE LAST AIRLINE IN THE WORLD TO OPERATE THE PASSENGER VERSION OF THE ICONIC McDONNELL DOUGLAS MD11 TRI-JET.

KLM Royal Dutch Airlines McDonnell Douglas MD-11
Flight KL763 from Amsterdam-Schiphol

CONTINENTAL AIRLINES HAS IN THE MEANTIME MERGED WITH RIVAL UNITED AIRLINES.
ALL THEIR JETS HAVE BEEN REPAINTED IN THE NEW UNITED LIVERY.

CONTINENTAL IS CAPTURED FOREVER IN THIS PHOTOGRAPH.

Continental Airlines Boeing 737-700
Flight CO818 from Newark-Liberty

#14

US Airways Boeing 757-200

Flight US1209 from Philadelphia International

US Airways Boeing 757-200
Flight US1207 from Philadelphia International

#63

#46

American Airlines Boeing 757-200
Flight AA667 from New York-John F. Kennedy

JetBlue Airways Airbus A320-200
Flight B6787 from New York-John F. Kennedy

THE PERFECT LIGHT FOR A SHOT WE HAD IN MIND FOR QUITE SOME TIME.

THE GORGEOUS PALM TREE WAS CUT DOWN SHORTLY AFTER.

KLM Royal Dutch Airlines Boeing 747-400
Flight KL785 from Amsterdam-Schiphol

#35

#25

Air France Airbus A340-300
Flight AF3510 from Paris-Charles de Gaulle

Air Canada Boeing 767-300ER
Flight AC1244 from Toronto-Pearson

#26

#09

American Airlines Boeing 737-800
Flight AA687 from Miami International

Air France Airbus A340-300
Flight AF488 from Paris-Charles de Gaulle

#04

#45

JetBlue Airways Airbus A320-200
Flight B6787 from New York-John F. Kennedy

United Airlines Airbus A320-200

Flight UA439 from Washington-Dulles

#38

#13

Corsair Boeing 747-400
Flight SS860 from Paris-Orly

US Airways Boeing 757-200
Flight US1562 from Charlotte-Douglas

#56

#21

Air Transat Airbus A310-300
Flight TS205 to Montreal-Trudeau

Air Canada Boeing 767-300ER
Flight AC1244 from Toronto-Pearson

#32

AT TAKE-OFF, PILOTS OF THESE INSEL AIR MD-80 SERIES JETS
LOVE TO GIVE FULL THRUST FOR SOME TIME, WHILE HOLDING THE BRAKES.

RESULTING IN THE LOUDEST MOMENTS ON MAHO BEACH, AS THE GROUND BEGINS TO QUAKE,
TRIGGERING ALARM SYSTEMS OF CARS PARKED NEARBY.

Insel Air McDonnell Douglas MD-82
Flight 7I521 from Curacao International

#49

Delta Air Lines Boeing 757-200
Flight DL440 to Atlanta Hartsfield-Jackson

Delta Air Lines Boeing 757-200

Flight DL439 from Atlanta Hartsfield-Jackson

#48

#12

American Airlines Boeing 757-200
Flight AA667 from New York-John F. Kennedy

Air France Airbus A340-300

Flight AF488 from Paris-Charles de Gaulle

#07

#83

US Airways Boeing 767-200ER

Flight US1209 from Philadelphia International

Continental Airlines Boeing 737-700
Flight CO818 from Newark-Liberty

#62

US Airways Boeing 757-200

Flight US1560 from Charlotte-Douglas

US Airways Boeing 757-200
Flight US1562 from Charlotte-Douglas

#55

JetBlue Airways Embraer 190
Flight B6450 from San Juan International

#69

#74

US Airways Boeing 757-200
Flight US1562 from Charlotte-Douglas

US Airways Boeing 757-200
Flight US1560 from Charlotte-Douglas

#64

#03

American Airlines Boeing 737-800
Flight AA687 from Miami International

Corsair Boeing 747-400

Flight SS860 from Paris-Orly

#53

Spirit Airlines Airbus A319-100

Flight NK100 from Fort Lauderdale-Hollywood

Continental Airlines Boeing 737-700
Flight CO818 from Newark-Liberty

US Airways Boeing 767-200ER
Flight US1209 from Philadelphia International

#42

#34

Delta Air Lines Boeing 737-700
Flight DL449 from Atlanta Hartsfield-Jackson

Corsair Boeing 747-400
Flight SS860 from Paris-Orly

#50

KLM Asia Boeing 747-400
Flight KL785 from Amsterdam-Schiphol

Copa Airlines Boeing 737-700
Flight CM228 from Panama-Tocumen

#71

#39

JetBlue Airbus A320-200

Flight B6787 from New York-John F. Kennedy

US Airways Boeing 757-200
Flight US1560 from Charlotte-Douglas

#73

#75

KLM Royal Dutch Airlines McDonnell Douglas MD-11
Flight KL763 from Amsterdam-Schiphol

Air Transat Airbus A310-300
Flight TS204 from Montreal-Trudeau

#65

#41

Delta Air Lines Boeing 737-700
Flight DL439 from Atlanta Hartsfield-Jackson

US Airways (Star Alliance) Boeing 757-200
Flight US1560 from Charlotte-Douglas

FUEL-GUZZLING QUAD-JETS LIKE THE BOEING 747 AND AIRBUS A340 ARE CURRENTLY BEING REPLACED BY MORE EFFICIENT TWIN-JETS ALL OVER THE WORLD.

THEIR FUTURE ON THE ST. MAARTEN ROUTES IS UNCERTAIN.

KLM Royal Dutch Airlines Boeing 747-400
Flight KL785 from Amsterdam-Schiphol

#02

A LIGHT IN THE BLACK.

LONG EXPOSURE OF A LATE ARRIVAL FROM MIAMI.

American Airlines Boeing 737-800
Flight AA729 from Miami International

#10

#15

JetBlue Airways Airbus A320-200

Flight B6787 from New York-John F. Kennedy

US Airways Boeing 757-200
Flight US1560 from Charlotte-Douglas

#08

#11

Continental Airlines Boeing 737-700
Flight CO861 from Newark-Liberty

United Airlines Airbus A320-200
Flight UA421 from Washington-Dulles

#59

KLM Royal Dutch Airlines Boeing 747-400
Flight KL785 from Amsterdam-Schiphol

#61

#40

Air France Airbus A340-300

Flight AF3515 to Paris-Charles de Gaulle

View of Maho Beach over wing, moments before departing to Paris
Seen from window seat 27K, aboard flight AF3515

#80

Copa Airlines Boeing 737-700
Flight CM229 to Panama-Tocumen

This first edition of *Jet Airliner: The Complete Works*
is limited to 2,000 casebound copies.



Most Press, Römerstr 104, 4600 Wels, Austria

Telephone & Facsimile: **43 7242 62829
Website: www.mostpress.com
e-mail: info@mostpress.com

ISBN 978-3-902600-11-0

Printed in China